Working Together for Safety

A State Team Approach to Preventing Occupational Injuries in Young People

Prepared by

Education Development Center, Inc.
55 Chapel Street
Newton, Massachusetts 02458–1060

Prepared for the National Institute for Occupational Safety and Health

DEPARTMENT OF HEALTH AND HUMAN SERVICES
Centers for Disease Control and Prevention
National Institute for Occupational Safety and Health

> This document is in the public domain and may be freely copied or reprinted.

Disclaimer

Mention of any company or product does not constitute endorsement by the National Institute for Occupational Safety and Health (NIOSH). In addition, citations to Web sites external to NIOSH do not constitute NIOSH endorsement of the sponsoring organizations or their programs or products. Furthermore, NIOSH is not responsible for the content of these Web sites.

Ordering Information

To receive documents or other information about occupational safety and health topics, contact NIOSH at

NIOSH
Publications Dissemination
4676 Columbia Parkway
Cincinnati, OH 45226–1998

Telephone: **1–800–35–NIOSH** (1–800–356–4674)
Fax: 513–533–8573
E-mail: pubstaft@cdc.gov

or visit the NIOSH Web site at **www.cdc.gov/niosh**

DHHS (NIOSH) Publication No. 2005–134

May 2005

SAFER • HEALTHIER • PEOPLE™

Acknowledgements

This report describes the Northeast Young Worker Resource Center, a project located at Education Development Center, Inc. (EDC), and funded (1998–2001) by the National Institute for Occupational Safety and Health (NIOSH), through award number U60/CCU116091–03. Raymond Sinclair is the NIOSH project officer. At EDC, Christine Miara is the project director and Susan Gallagher the principal investigator. This publication was written by Marc Posner, with graphic design by Ronnie DiComo, both of EDC.

We wish to thank the members of the State teams and community projects upon whose work this publication is based and who provided documents and materials, allowed themselves to be interviewed, and reviewed drafts of the manuscript. Anne Stirnkorb, Vanessa Becks, and Susan Afanuh provided editorial and graphic assistance.

Contents

Introduction . . . 1
　History of Young Worker State Teams in the Northeast . . . 1
　Northeast Young Worker Resource Center . . . 2
　Overview . . . 3

State Teams in Action: Two Case Studies . . . 5
　The New Hampshire Teen Workplace Safety Coalition . . . 5
　The Connecticut Young Worker Team . . . 8

Working as a State Team . . . 11
　Working at the State Level . . . 11
　Working in the Community . . . 13
　Working Regionally . . . 13

Taking Action: Strategies and Projects to Prevent Injuries to Young Workers . . . 15
　Curricula and Safety Training for Youth . . . 15
　Other Ways of Educating Teens . . . 17
　Working with Educators . . . 19
　Educating Parents . . . 22
　Working with Employers . . . 22
　Educating Health Care Providers . . . 25
　Using Data . . . 26
　Providing Information about the Law . . . 27
　Evaluation and Program Monitoring . . . 29
　Systematic Reform . . . 29

Resources . . . 33
　Young Worker Safety Resource Center . . . 33
　State and Regional Contacts . . . 34
　Resource Materials . . . 40

Introduction

Most young people work at some time during high school. Although working can be a positive experience, it also has risks. The Institute of Medicine's Committee on the Health and Safety Implications of Child Labor reports that 50 percent of youths between ages 15 and 17 work at some time during the course of a year and that 80 percent of students work at least some time during high school. Every year, at least 100,000 of these young people seek treatment in an emergency room for a work-related injury. Every year, at least 70 young people are killed on the job. Young people are injured in the workplace at twice the rate of adult workers. Yet no single agency has the ultimate responsibility for protecting young people from workplace hazards. What is needed is an approach that brings coherence and coordination to this mission.

A State team for young worker safety is a coalition of agencies and organizations whose goal is to protect the safety and health of young people in the workplace. The American Heritage Dictionary defines a team as "a group organized to work together." This definition goes to the heart of the State team approach. A State team is not a committee, taskforce, or blue ribbon panel. State teams do not exist to make recommendations, issue reports, share information, or discuss issues—although they can do all of these. State teams exist to work on concrete projects that protect young people from injuries in the workplace. Over the past 5 years, several of the States in the Northeastern part of the United States have successfully used the State team approach to improve their capacity to protect young workers.

History of Young Worker State Teams in the Northeast

The efforts described in this publication build on the work of three community projects funded by the National Institute for Occupational Safety and Health (NIOSH) in 1995:

- ✓ The Labor Occupational Safety and Health (LOSH) Program at the University of California at Los Angeles worked with the schools in that city.

- ✓ The Labor Occupational Health Program (LOHP) at University of California at Berkeley implemented a program in nearby Oakland, California.

- ✓ The Massachusetts Department of Public Health, working with Education Development Center (EDC), undertook a project in Brockton, Massachusetts.

For the next 3 years, these projects developed curricula for high school students, materials for employers and parents, strategies for providing effective training, and other resources and strategies. The grantees piloted these materials and strategies in the three communities.

Promoting Safe Work for Young Workers, a publication available from NIOSH, contains a detailed description of these projects.

In 1998, the Educational Development Center (EDC) invited staff from departments of health, education, and labor in the six New England States, New York, and New Jersey, along with representatives from NIOSH, to a meeting on how the experiences of the three pilot projects could benefit young workers in the Northeast. Participants explored how agencies and organizations could work together to prevent injuries to teen workers. It became apparent that although each agency played an important role in preventing these injuries, no single agency had a clear responsibility to address all aspects of young worker safety. By the end of the day, teams from each State had begun outlining plans for coordinated efforts to prevent injuries to young workers.

Northeast Young Worker Resource Center

NIOSH staff attending the meeting at EDC recognized that the State teams presented an important way of building State capacity to protect young workers. Thus, NIOSH funded the creation of the Northeast Young Worker Resource Center (NYWRC) at EDC. The purpose of the NYWRC was to facilitate the development and ongoing work of the State teams with technical assistance and resources, and thus to bring the experience of the three pilot projects to a broader audience. The assistance and resources provided by the NYWRC to the State teams and their community projects included the following:

- ✓ An annual training meeting

- ✓ Training in the use of young worker safety curricula for teachers, job placement professionals, and State agency staff

- ✓ Periodic resource mailings, which included materials developed by State teams

- ✓ Adaptations of two occupational safety and health curricula for high school students to reflect the data, child labor laws, and resources of individual States

- ✓ Adaptations of brochures originally created for Massachusetts parents, employers, health care providers, and teens to reflect the data, child labor laws, and resources of individual States

- ✓ Technical assistance through site visits, telephone conversations, and e-mail
- ✓ A regional teen occupational safety electronic discussion list
- ✓ Support for the Northeast Young Worker Network, the State teams' membership organization

Overview

Working Together for Safety: A State Team Approach to Preventing Occupational Injuries to Young People was developed by EDC with funding from NIOSH. It begins with two case studies that demonstrate the value of the State team approach. The remainder of the document describes the experiences and activities of the State teams in the Northeast; the products developed by the teams for teens, parents, employers, school staff, health care providers, and others who can help protect young people from injury on the job; and key resources for other States interested in creating their own State teams.

This publication reflects the hard work of a large number of dedicated people. We hope you will take advantage of their efforts and experience by using the strategies and activities described in these pages to prevent injuries to working youth in your State.

Note: No Federal funds were used to support or participate in any of the advocacy or legislative activities in any of the States.

State Teams in Action: Two Case Studies

The New Hampshire Teen Workplace Safety Coalition: A Case Study in Team-Building While Working Toward a Long-Range Goal

The experience of the New Hampshire Teen Workplace Safety Coalition demonstrates how a State team can

- ✔ take advantage of the strengths and resources of each of its members and
- ✔ work on manageable short-term activities while maintaining a focus on a longer-range goal.

Getting Started

In the summer of 1998, the New Hampshire legislature abolished the work permit requirement for workers aged 16 and 17. This action took place with little warning during the summer, when schools were out of session. Many of the people in New Hampshire's Departments of Health, Education, and Labor who would become active in the New Hampshire Teen Workplace Safety Coalition learned about this legislation only after it was passed.

This event provided an immediate impetus for the members of the nascent State team, who were preparing to attend the initial young worker safety meeting at EDC. Lynda Thistle-Elliot of the New Hampshire Department of Education reported, "The repeal of the work certificate system showed us that we needed some coordination. We had people who were working on teen worker safety. Yet this slipped by." The team understood that reinstating work certificates was not immediately possible. However, they found that they could do a lot to protect young workers. Representatives from the Departments of Health and Education volunteered to chair meetings; team members took turns taking minutes; and all members became involved in the team's projects.

CD-ROM
(Sugar Valley Regional Technical Center, New Hampshire)

Creating Educational Materials

One of the State team's first projects was to develop a fact sheet for parents. Although this was a relatively quick activity, it gave team members the chance to focus as a group on issues related to young worker safety and to learn firsthand what each member's agency and organization had to offer. The team used data from the Departments of Health and Labor and child labor law information from the Department of Labor and the Safety Council to create *Working Teens—A Guide for Parents*. The Safety Council and Department of Labor paid for printing the fact sheet, while the Department of Education distributed copies to every school district in the State.

The team's next project was to create and distribute book covers featuring information about child labor laws. A number of the agencies and organizations represented on the State team contributed resources or capabilities to help this project succeed. The Department of Education sponsored a contest in which high school students designed book covers with a young worker safety motif. The State team selected the winning design. The Department of Labor checked the information on the book covers for accuracy. The New Hampshire Municipal Association (a membership organization of cities, towns, and townships), State Department of Labor, and the Occupational Safety and Health Administration (OSHA) funded the printing of more than 30,000 book covers, which were distributed to students by the Department of Education. The contest generated considerable media attention and helped raise public awareness about workplace safety for teens. Lynda Thistle-Elliot reported that having "an actual product on which to work generated enthusiasm. It kept us going."

Supporting a Community Project

Another important activity was the State team's support of a community-based project and subsequent use of the results to benefit other communities. The New Hampshire State team awarded its NIOSH-funded mini-grant to Sugar River Valley Regional Technical Center, a regional vocational education high school. Staff and students at the school used this grant, and information provided by the coalition, to create a CD-ROM entitled Teen Health and Safety in the Workplace. Sugar River Valley, and the State's other regional technical centers, now use this CD-ROM, along with supplemental print materials and teacher training, to teach teens about workplace safety. The materials are also used by State team members in presentations to educators, State agency personnel, and youths.

Expanding the Team

The benefits of having key organizations represented on the team became evident as the team began to reach out to new audiences. The representative from the New Hampshire chapter of the National Safety Council, for example, arranged for State team members to speak and distribute information at that group's annual conference.

The team also learned that opportunities can be missed if key State agencies are not represented. The State team discovered that the State Department of Labor's Wage and Hour Division had published an article on young workers in New Hampshire Business Week that overlooked the issue of safety, and had also created a bookmark containing information about State child labor laws that neglected to mention that some of these laws were superseded by stricter Federal law. The team recruited Wage and Hour representatives to ensure that such opportunities would not be lost in the future.

The coalition did not let its projects, team-building, and expansion overshadow its initial concern. In spring 2001, two State legislators introduced legislation reinstating the work permit requirement. This time, team members were prepared to reach out to schools, parents, and employers with information about work certificates. The coalition has done much to raise awareness of occupational injuries to teens, and continues to make New Hampshire a safer place for young people to work.

New Hampshire Teen Workplace Safety Coalition Members

The New Hampshire State team includes representatives from the following groups:

- United States Department of Labor
  — Wage and Hour Division, Manchester District Office
  — Occupational Safety and Health Administration (OSHA), Concord Area Office
- New Hampshire Department of Education
  — Bureau of Integrated Programs
  — School-to-Work
- New Hampshire Department of Health and Human Services
  — Injury Prevention Program
  — Occupational Health Program
- New Hampshire Department of Labor
  — Inspection Division
  — Wage and Hour Administration
- Schools
  — Portsmouth High School
  — Sugar River Valley Regional Technical Center
- Other Agencies and Organizations
  — Emergency Department, Concord Hospital
  — Labor Committee, New Hampshire House of Representatives
  — Injury Prevention Center, Dartmouth Medical School
  — Jobs for N.H. Graduates, Inc.
  — New Hampshire Committee on Occupational Safety and Health (COSH)
  — New Hampshire School Boards Insurance Trust
  — Safety and Health Council of New Hampshire

The Connecticut Young Worker Team: A Case Study of Young Worker Safety Training

The Connecticut Young Worker Team demonstrates how agencies and organizations working together can have a far greater impact than any one group could produce on its own. The Connecticut State team did the following:

- ✔ Brought together professionals, each with the responsibility for one aspect of young worker safety. They had never worked together and, in many cases, had not even met one another, despite their common interests and responsibilities.

- ✔ Established multidisciplinary training teams

- ✔ Provided safety training in a number of different venues

- ✔ Integrated safety training into a variety of education, health, and job training systems

Connecticut Young Worker Team Members

The Connecticut State team includes representatives from the following:

- United States Department of Labor
  — Occupational Safety and Health Administration
  — Wage and Hour Division
- Connecticut Department of Education
  — Bureau of Career and Adult Education
- Connecticut Department of Labor
  — Occupational Safety and Health Administration
  — Quality Program Review
  — Wage and Workplace Standards Division
- Connecticut Department of Public Health
  — School and Primary Health Unit
  — Injury Prevention Program
  — Environmental Epidemiology and Occupational Health
- Connecticut Workers' Compensation Commission
- Workforce Investment Boards
  — Capitol Region Workforce Development Board
- Other Agencies and Organizations
  — ConnectiCOSH (Committee for Occupational Safety and Health)
  — Middletown Health Department
  — Middletown High School
  — Governor's Career Internship Partnership
  — Community Enterprises, Inc.

Getting Started

Some of the key participants did not have any experience with young worker safety, much less organizing a multidisciplinary coalition to address this issue. Marian Storch of the Connecticut Department of Public Health's Injury Prevention Program, a founding member of the team, reported, "It took us awhile to focus. The reason the health department got involved was because of the training on youth occupational safety held at EDC. Prior to that, young worker safety wasn't even on our radar screen. We didn't even know who else in our department was a natural partner. After the forum at EDC, we discovered that this was a topic of interest across several divisions in the department."

After several meetings focusing on the health implications of youth employment, team members began to reach out to people they knew in other agencies. The team expanded to include representatives from (1) several divisions of the State Departments of Health, Education, and Labor, (2) the regional office of the United States Department of Labor, a local workforce development board and local health department, (3) COSH, and (4) the Governor's Career Internship Partnership. A broad representation of State agencies and offices proved key to the team's ability to use those systems with access to working youth and employers Team members share the duties of chairing meetings and sending out followup notices.

Supporting a Community Project

Awarding a mini-grant to a community project (using NIOSH funds) provided the team with an initial focus, as well as an important lesson about integrating safety training into programs that connect youth to the workplace. The mini-grant project was coordinated by the Middletown Health Department, which also administers the city's summer jobs program. The Health Department hired Anita Vallee, a high school teacher, to provide a three-hour training, using *Work Safe!* (a curriculum developed by the Labor Occupational Health Program at the University of California at Berkeley), to youth in the Middletown summer jobs program.

The activity was very well received by teens and their supervisors in the job training program. Vallee later incorporated *Work Safe!* into career awareness classes at Middletown High School, trained other high school business and career education teachers to use the curriculum, and expanded the summer job program training project. She also surveyed local employers to identify specific work hazards and issues that should be emphasized in the program. This project demonstrated the utility of schools, job training programs, and workplaces as venues for young worker safety training, as well as the value of using a brief yet engaging and effective curriculum like *Work Safe!*

Conducting Outreach and Training

Most members of the Connecticut State team were trained to use *Work Safe!* by the NYWRC. In addition to delivering the full three-hour training, team members also provide abridged training (for events lacking time for a complete training) and shorter presentations about the curriculum and young worker safety at professional meetings, workshops, and other events. The active

Newspaper article (Connecticut)

training and presentation schedule maintained by the team resulted from a decision to target teachers and job placement professionals who have opportunities to use the curriculum with young people. Marian Storch reports, "We are trying to knock on as many doors as we can reach. We'd like to get young worker safety training institutionalized in as many different settings as we can." This strategy is proving its worth. In December 2000, for example, Jennifer Stefanik, a team member from the Capitol Region Workforce Development Board, convinced the board to require State-funded youth employment programs to incorporate young worker safety training into their programs. Stefanik teaches the supervisors and directors of youth programs to use *Work Safe!*; they, in turn, use the curriculum in their own training for the youth they employ. The Connecticut Young Worker Team is in the process of expanding this program to all eight Workforce Development Boards in the State.

Finding New Audiences

The Connecticut Young Worker Team used existing meetings, conferences, and training events as opportunities to introduce teen worker safety to new audiences, such as the following:

- Annual Cooperative Work Education/School-to-Career Joint Conference
- Connecticut Association for Adult and Continuing Education Annual Conference
- Connecticut Department of Labor Youth Forum
- Annual Meeting of the Connecticut Association of Family and Consumer Science Educators
- Connecticut Association of Work-Based Learning Fall Conference
- Annual Connecticut School-to-Career Summer Institute
- Connecticut Department of Corrections Vocational Education Instructors
- Bridgeport Department of Education (Technical Education, Aquaculture, and Family and Consumer Science Departments)

Working as a State Team

The State teams, assisted by the NYWRC, work at three levels: State, community, and regional.

Working at the State Level

Each team worked to organize systems and take action to improve young worker safety in the State as a whole. The first step in this process was to recruit additional members. Most teams started with one or two members who called their counterparts in other agencies and organizations for an initial meeting. The teams expanded through personal contacts and by recognizing the agencies and organizations in each State that were critical to the team's mission. This provided each team with more resources and access to those systems and audiences essential to young worker safety.

A variety of organizations and agencies can make unique contributions to the work of the team:

- ✓ The State department of labor can provide knowledge about child labor laws, health and safety laws, data on injuries and labor law violations, and funding.

- ✓ The State department of education can contribute access to school personnel and students, educational expertise, and a perspective on integrating young worker safety activities into school programs.

- ✓ The State department of health can offer access to health care providers, experience with health education and adolescent health, data on work injuries, and information about medical care for these injuries.

- ✓ Committees on Occupational Safety and Health (COSH) can work with other labor organizations and provide expertise on occupational health and safety.

- ✓ National Safety Council chapters can provide access to employers as well as funds and other resources.

✓ Employers, job trainers, and others who work directly with young people can provide important expertise as well as access to worksites in which young worker safety training can take place.

In each State, a lead agency or organization volunteered to coordinate the team's efforts. The teams quickly discovered that having manageable tasks and objectives was important to the teams' progress. Dividing tasks into manageable parts (for example, looking at data, recruiting new members, providing training and presentations, and scheduling and facilitating meetings) and assigning each to a different member allowed the team to make progress without unduly burdening any one member or agency.

Teams also discovered that starting with a project that provides an immediate sense of accomplishment is important. Small projects with a concrete product provide the momentum and enthusiasm teams need to persevere and work toward more ambitious goals. These projects included producing brochures on teen worker safety for health care providers and presenting information about young worker safety at meetings of employers, educators, and other professional associations.

Who Should be Represented on a State Team?

Agencies, organizations, and disciplines represented on State teams in the Northeast include the following:

- State and Federal agencies
  - State departments of labor (wage and hour, workers' compensation, occupational safety and health, and education divisions)
  - State departments of education (School-to-Work and vocational education offices)
  - State departments of health (injury prevention, adolescent health, and occupational health programs)
  - Workforce investment boards (and the State office that oversees the boards)
  - Regional offices of Federal OSHA and the Wage and Hour Division of the United States Department of Labor.

- Persons who work directly with young people
  - Business and vocational education teachers
  - Job training program staff
  - Youth development program operators
  - Employers

- Representatives from organizations with an expertise and interest in occupational safety or adolescent health
  - COSH
  - Labor unions
  - Safety councils, representing businesses in the State
  - Chambers of commerce
  - Hospitals and health care organizations
  - State school board insurance providers
  - Safe Kids Coalitions, Safe Communities coalitions, and other community-based injury prevention programs
  - University departments of workplace environment and occupational safety

Working in the Community

Community projects that work directly with youth provide a sense that the team is making a difference. This can sustain the team over the longer period needed to institutionalize young worker safety training in job programs, implement curricula in schools, or change State labor laws. An enthusiastic teacher, public health professional, or youth safety advocate can do much with a small grant, some young worker safety materials, and a bit of technical assistance. Each State team recruited and helped a community agency or organization implement an educational project to prevent injuries to young workers. NYWRC provided technical assistance and $2,500 stipends to each community project. The State teams created their own criteria for awarding these stipends. They assisted the community projects and often found ways of using the strategies and materials created by these projects in other communities in the State (and the region). A community project that generated community support and demonstrated effectiveness could provide the foundation for implementing similar projects in other communities or across the State, or replicating the project in other States in the region.

Working Regionally

With support from the NYWRC, the State teams established the Northeast Young Worker Network. Resource Center staff brought the teams together once a year and shared information, ideas, resources, and materials throughout the year. This networking benefits individual States and the region as a whole in a variety of ways. Brochures and other materials created by one State or community are often adapted for use in other States or communities. Impending Federal legislative or rule changes that escape the notice of one State team may be caught by another. Activities in one State sometimes catalyze action in its neighbors. And the regional meetings demonstrate to others a widespread and active interest in young worker safety.

The Power of Regional Coordination

Christine Miara coordinates the Young Worker Safety Resource Center. Here, she speaks to the value of having representatives from all the young worker State teams meet once a year at a networking and training event:

"Robin Dewey [our training consultant from the University of California at Berkeley Labor Occupational Health Program] and I had several goals for our annual regional meetings. We wanted to give participants the opportunity to showcase educational resources they had developed over the past year. We wanted to provide new research and information about national initiatives. And we hoped that by allowing people from each State and community to describe their activities, those in the other States would be motivated to expand the scope of their work.

"Based on evaluations and participant feedback, it would seem that we met these goals. For example, at the end of the meeting at which New Hampshire shared its young worker CD-ROM and book cover, many people from other States said that they planned to adapt some of these ideas and materials.

"At another meeting, the head of the United States Department of Labor's Child Labor and Special Employment Team described effective ways to strengthen and enforce child labor laws, and several States discussed specific child labor restrictions they had passed. Hearing from their counterparts in other States and having the issue put in a national context encouraged members of several State teams to move beyond a purely educational approach to young worker safety.

"People who are relatively new to an issue, such as young worker safety, will be much more motivated and able to undertake a broad range of activities if they can come together to learn from, and hold discussions with, experts and colleagues. Regional collaboration also helps States avoid duplication of efforts and maximize the use of limited resources."

Taking Action: Strategies and Projects to Prevent Injuries to Young Workers

A variety of strategies and projects can be used to prevent injuries to young workers. The State teams in the Northeast teach teens about workplace safety; educate parents, employers, health care providers, educators, and government officials; collect and analyze data; and enhance and enforce child labor laws.

Meeting the Challenge: Obtaining Materials for Your State

Many of the curricula, brochures, and other products developed or used by the northeastern State teams can be adapted for your State. For more information, contact the relevant State team or agency or EDC. NIOSH funding for the NYWRC ended in October 2001. EDC and the Labor Occupational Health Program have received an OSHA grant to create the National Young Worker Safety Resource Center, which provides training and resources to States working on the issue of teen safety in the workplace. Resource center staff may be able to adapt some of the curricula and brochures mentioned in this section to reflect the child labor laws in your State. Information about contacting the National Young Worker Safety Resource Center can be found in the Resources section of this publication.

Brochure (Young Worker Safety Resource Center)

Curricula and Safety Training for Youth

Young worker safety curricula teach teenagers to recognize their right to a safe working environment and to become safer workers. Teenagers need to be taught how to work safely. They should be aware that State and Federal laws protect them in the workplace. And they need to know how to identify hazards and negotiate with employers over unsafe conditions and to whom they can turn if these negotiations are unsuccessful. These issues are covered in three curricula; two that were developed with NIOSH funding, and a third that was modeled after them:

Curriculum (Massachusetts/ Education Development Center, Inc.)

✓ *Safe Work/Safe Workers* (created by EDC and the Massachusetts Department of Public Health) is used in Rhode Island, Massachusetts, New Jersey, New York, New Hampshire, and Vermont.

✓ *Work Safe!* (created by the Labor Occupational Health Program at the University of California at Berkeley) is used in an expanding number of schools in Connecticut.

✓ Maine's curriculum, *Starting Safely*, is based on these curricula.

States in the Northeast use these and other materials in a number of settings, including schools and job training programs.

A number of groups have incorporated safety curricula or training into their programs:

Curriculum (Maine)

✓ **School-to-Career programs:** The Bloomfield (New Jersey) Health Career Foundation augmented *Safe Work/Safe Workers* with information about risks to health care workers (such as bloodborne pathogens). With funding from the New Jersey Department of Labor, the foundation trained high school teachers and administrators involved with the Health Science Career Program to use the modified curriculum.

Other teachers in New Jersey's School-to-Career and vocational education programs use *Safe Work/Safe Workers* in work-readiness courses. Students who complete the curriculum and pass a test receive a Safety Skills Certificate from the New Jersey Safety Council. School-to-Career staff report that it is easier to place students who have earned certificates because employers believe they are less likely to be injured on the job.

Safety certificate checklist (Maine)

✓ **Health education and other academic classes:** The Maine team is working with that State's Coordinated School Health Project to include occupational safety and health in school health curricula. The Coordinated School Health Project is part of a Federally funded initiative to help State departments of health and education develop school health curricula.

- ✓ **Vocational education programs:** Maine allows vocational education students to earn a Safety Certificate after completing 55 hours of coursework and passing written and performance examinations. The coursework can be integrated into specific vocational classes or taught as a stand-alone program.

- ✓ **Job programs:** The Middletown, Connecticut, summer jobs program uses *Work Safe!* to prepare youth for summer employment. In Maine, *Starting Safely* is used in a year-round job training program for at-risk youth.

- ✓ **Workforce Development Boards:** The Connecticut State team is working with that State's workforce development boards to put into place a requirement that the youth job training programs funded by the boards use *Work Safe!* to teach their young participants about workplace safety. The State team conducts regional training in which both members of the workforce development boards and youth job training program operators learn how to use the curriculum.

> **Meeting the Challenge: Getting into the Schools**
>
> In addition to teaching the "three R's," schools are now expected to teach students to resolve conflicts, master new technologies, and resist peer pressure to use drugs and alcohol. Lynne Lamstein of the Maine Department of Labor describes getting teachers to use *Starting Safely*, Maine's young worker curriculum, as an "ongoing challenge." She reports, "We have learned to run with the opportunities we get. We are most successful with vocational education staff and school-to-work educators. Our greatest achievement in this area came when Jobs for Maine's Graduates began requiring that their teachers use *Starting Safely*. We have also outlined how the *Starting Safely* aligns with the Maine Learning Results. This allows our program to be seen as an integral part of the curriculum, rather than as an 'add-on.'"
>
> Jobs for Maine's Graduates is a State affiliate of Jobs for America's Graduates, an initiative under which States help disadvantaged youths complete high school, enroll in college or vocational school, and find and keep jobs

Other Ways of Educating Teens

The most comprehensive method of teaching teens about workplace safety is with a young worker safety curricula. However, other strategies require less time and can reach larger numbers of youths. These activities can help build momentum in States or communities not yet ready to implement a safety curriculum and can reinforce safety messages in those that do. Reaching teens is also a good way to reach their parents and help build public support for young worker safety. The northeastern State teams found a number of ways to provide safety information to teens:

Teen booklet (Massachusetts)

Book cover
(New Hampshire)

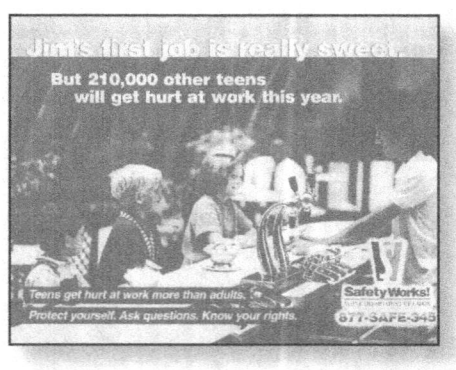

Poster
(Maine)

✓ **Booklets:** The Massachusetts Department of Public Health (in collaboration with the Massachusetts Office of the Attorney General) produced *Do You Work? Protect Your Health, Know Your Rights. A Guide for Working Teens*, a 12-page illustrated booklet describing the protection child labor laws offer teens and whom they can contact if they have questions about safety, health, or work hours. The New Jersey Department of Labor prepared a similar booklet.

✓ **Book covers:** The New Hampshire State team sponsored a contest in which high school students designed book covers featuring information about child labor laws. The contest generated considerable media attention and a substantial level of awareness among students, parents, and school staff. The Department of Education sent the book covers to all schools in the State.

✓ **Wallet cards, posters, and mousepads:** The Maine Department of Labor provides educators with copies of *Rights for Working Teens*, a wallet card that is distributed with work permits. Maine produced a color poster informing teens about their rights in the workplace and whom to call for more information. Such materials can be distributed at school, during career fairs and orientations for job shadow days, and almost anyplace teens gather, including malls, fairs, concerts, and parks.

✓ **Public Service Announcements (PSAs):** The Maine Department of Labor produced PSAs that were shown in movie theatres (at a special nonprofit rate). Other theatre chains around the country have agreed to show young worker safety PSAs without charge. Maine also produced two 30-second PSAs for television on teen worker safety that have been aired across the State.

✓ **CD-ROMs:** A mini-grant project funded through the New Hampshire State team developed a young worker safety CD-ROM, which students in vocational education schools use to teach themselves about occupational safety.

✓ **Presentations:** Several State teams introduce teenagers to occupational safety through short presentations in vocational education, career, and business classes. This strategy could also be used during assemblies and orientation sessions for summer job or recreation programs.

- ✓ **Teen focus groups:** The Maine Department of Education used focus groups to explore what teenagers know about workplace safety, where they came across this information, and how likely they are to speak up about workplace hazards. Many interesting findings emerged: (1) teens knew little about the child labor laws and viewed them more as restrictions than rights, (2) the level of supervision among teens varied greatly, and (3) their primary safety concern was late-night security. The information gleaned from these groups informs Maine's development of educational materials for teens and employers.

- ✓ **Child labor law calendar:** The Connecticut State team, in collaboration with a middle school business class, designed a calendar featuring that State's child labor laws. The calendar was printed with funds from the Connecticut Department of Labor Wage and Workplace Standards Division and distributed State-wide.

- ✓ **Teen peer education and assessment project:** The Massachusetts Committee for Occupational Safety and Health (MassCOSH) is involved in a teen peer education and assessment project funded by a mini-grant from the State team. MassCOSH recruited a small group of teenagers ranging in age from 12 to 16. These teens were introduced to occupational health and safety and child labor laws in a 4-hour training, during which they developed a peer worker survey (based on the surveys used by EDC and other States). The survey explores teens' knowledge about occupational safety, child labor laws, and workplace rights; their experience with occupational injuries; and the types of young worker safety programs that appeal to teenagers.

Calendar
(Connecticut)

The teenagers participating in the project used the survey to collect information from their peers in the community. The results were presented to a meeting of community youth groups and in a report that informed the development of an occupational health and safety peer education program.

Teen booklet
(New Jersey)

Working with Educators

Educators are essential partners for any comprehensive effort to teach young people about occupational safety. But few educators have received adequate training about occupational injuries and their prevention. Educating teachers, guidance counselors, and other school staff is an important step in reaching students with young worker safety curricula and other materials, as well as ensuring that schools are making the best use of systems already in place (such as the work permit process) to protect young people. State teams have brought this information to educators in a variety of ways:

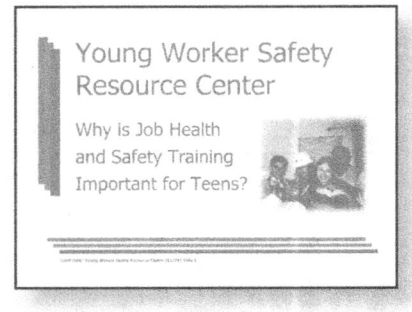

Slide show
(Young Worker Safety Resource Center)

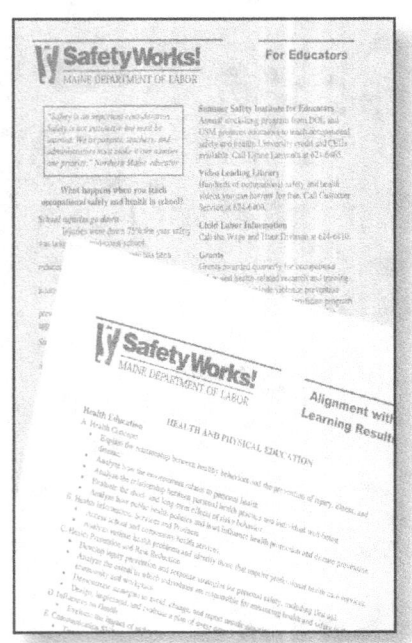

Teacher resources
(Maine)

✓ **Conferences:** A Maine State team representative spoke and exhibited information about occupational health and safety at health education conferences for teens, school nurses, school wellness staff, and principals. Most health teachers had not been trained in occupational safety and were unaware of the risks their students faced on the job. Connecticut has provided abbreviated *Work Safe!* training-of-trainers at schools and vocational education conferences, often using the *Why Is Job Training in Safety Education Important to Teens?* slide show developed by EDC and the Labor Occupational Health Program under OSHA funding.

✓ **Professional development:** The Connecticut Department of Education offers training on *Work Safe!* at its annual School-to-Career Summer Institute. The New Jersey Department of Labor funded training on a modified version of *Safe Work/ Safe Workers* (with more specific information about risks to health care workers) for high school teachers and administrators participating in the Health Science Career Program.

The Maine Department of Labor and the University of Southern Maine sponsors an annual week-long Summer Safety Institute for Educators at which teachers are trained to teach occupational safety and health.

✓ **Flyers:** The Maine Department of Labor produced a simple, easily photocopied, one-page flyer for educators on why occupational safety and health should be taught in schools and the services the department can provide to schools. They also created a flyer explaining how their young worker safety curriculum aligns with the learning results required of Maine's health education, physical education, and career education programs.

- ✓ **Department of Education mailing:** The Massachusetts Commissioner of Education sent a letter to every superintendent in the State informing them that young worker safety brochures for teens, parents, and employers, as well as training on young worker safety curricula, were available from the Department of Public Health. This mailing resulted in requests for almost 50,000 copies of the brochures, which the Department of Education printed and distributed, as well as about 30 requests for training. This experience demonstrated to the Department of Education that there was a demand for young worker safety resources.

- ✓ **Work permit guidelines:** The granting of work permits by schools is often left to school staff who have little or no knowledge of the State or Federal child labor laws. This is a missed opportunity to educate teens and parents at the very point that teens are entering the job market. The Massachusetts State team is one of several teams exploring the possibility of developing guidelines that would clarify the work permit system for schools and use the process for providing key information on safety and child labor laws to teens, their parents, and their employers.

Parent booklet
(Massachusetts)

Meeting the Challenge: Recruiting Educators as Partners

Judith Andrews of the Connecticut State Department of Education relates how she leapt at the chance to join the State team:

"A year before we officially became a State-wide youth safety training team, Sue Prichard at the Connecticut Department of Labor told me that they had been approached by the Connecticut Department of Health to develop a youth safety curriculum. I asked to be kept in the loop. I had known since first reading the Federal School-to-Work legislation in 1994 that addressing youth safety was first and foremost the responsibility of educators. But I never had much luck selling this idea. Educators initially thought that anything work-related had to be the responsibility of employers. When Sue called, we at the State Department of Education were already beginning to focus more on work-based learning because of our School-to-Career Initiative and cooperative work education program. This project offered a curriculum-based approach with which educators could connect. We jumped right on the train. It was perfect."

Educating Parents

Parents have substantial control over the work lives of their children. Teens usually require parental permission to work—if not legally, at least within the family. Parents need to be aware of the dangers their children might face on the job—dangers not only to their health, but also to their academic performance (and, thus, their future). Parents need to learn about safety issues that may not have been in the public consciousness during their youth, such as repetitive motion injuries. Parents also need to know that laws are in place to protect their children on the job. And parents need to understand the ways in which they can help protect their children. Because parents are also employers, educators, health providers, government officials, and voters, educating parents can help build public and political support for young worker safety.

One way in which State teams reach parents is through booklets and fact sheets. For example, the Massachusetts Department of Public Health, in collaboration with EDC, developed *Protecting Your Working Teen: A Guide for Parents*. With assistance from EDC, the Maine State team adapted this booklet for that State's laws and resources. These booklets are sent home with schoolchildren, distributed at malls, health fairs, neighborhood festivals, health care facilities, and during school open-houses and parent-teacher conferences.

In another example, the New Hampshire State team used the information in Massachusetts' *Protecting Your Working Teen* to create *Working Teens: A Guide for Parents*. This fact sheet contains data, child labor law information, and contacts specific to New Hampshire. It was designed to be photocopied on two sides of a single letter-size piece of paper and is thus less expensive to duplicate than the original booklet.

Working with Employers

Employers do not want young people to be injured in their workplaces. In addition to their concern for the health of the teens they employ, injuries are bad business. Minimally, an injured employee cannot work—or cannot work to full capacity. Injuries can provoke conflicts with parents, insurance companies, OSHA, and State labor departments and may result in fines, lawsuits, increased workers compensation premiums, and bad

Parent fact sheet
(New Hampshire)

publicity. State teams have worked with local employers, franchises, regional corporate offices, and organizations to which employers belong such as retail, business, or trade associations, chambers of commerce, or Rotary Clubs. A number of methods have helped turn employers into allies:

✔ **Focus groups:** The Maine Department of Labor held focus groups for representatives of companies that employ young people to learn about the types of jobs young people do, whether they are given safety training on the job, and what the department can do to help employers prevent injuries to young workers. Some interesting findings emerged, including the following:
(1) employers do not have in-depth knowledge of the child labor laws, (2) teen workers are given the same training as adults, and (3) employers need activity-based training and would welcome resources and collaboration.

✔ **Workshops and training:** The Massachusetts State team and its mini-grant project at the University of Massachusetts at Lowell held a workshop for employers. The participants were primarily from retail stores and restaurants. Topics included the child labor laws, industry-specific safety hazards, the work permit system, and how teens differ from adults in ways that affect their safety on the job. Employers also had the opportunity to present their questions and concerns to representatives of the State team and to participate in a hazard-mapping exercise to enhance their ability to identify health risks in the workplace.

The Massachusetts State team and the University of Massachusetts at Lowell also facilitated a session on young worker safety as part of a breakfast series hosted by the local Council on Regional Economic Development. Inviting leaders from youth-serving organizations (such as the Boys and Girls Club and the YMCA) helped connect the issues of economic development and youth development.

Massachusetts data revealed that burn injuries to teen workers were a special problem in some retail food establishments. Many of these burns were associated with coffeemakers. Members of the Massachusetts State team then held training for 450 retail bakery franchise owners and managers. Presentations were made on child labor laws,

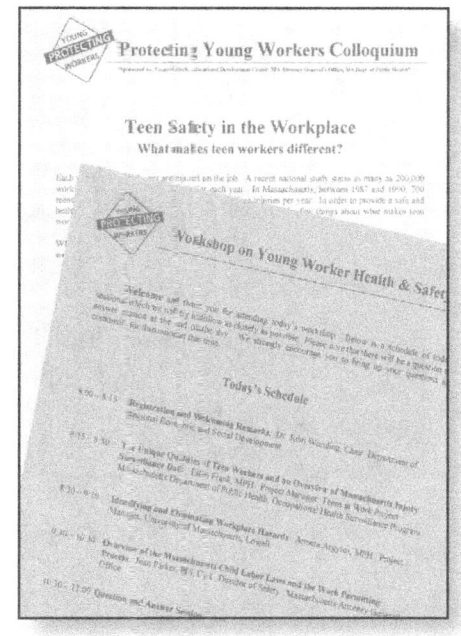

Employer workshop materials
(Massachusetts)

Employer kit
(Maine)

Employer guide
(Massachusetts)

injuries to young workers, and prevention strategies. The brew baskets of the coffeemakers are being redesigned to reduce burn injuries.

✓ **Employer conferences and meetings:** Members of the Maine, New Jersey, and New Hampshire State teams present and exhibit information about teen worker safety at the annual safety and health conferences in their States.

✓ **Training and educational materials for employers:** The Maine Department of Labor used funds from a child labor violation settlement to create *SAFETEEN*, a kit that helps employers teach their teenage employees about workplace safety. Each *SAFETEEN* kit includes a book of safety training exercises that employers can use to teach teens about workplace safety, a booklet for employers about workplace safety and child labor laws, multiple copies of a booklet for teens about workplace safety, wallet cards, posters, and information about obtaining *Starting Safely*, Maine's young worker safety curriculum. The kit was sent to employers under a cover letter signed by Maine's governor and is distributed through Maine's Career Center network. Maine is currently evaluating this effort.

The Massachusetts Department of Public Health published *Young Worker Health and Safety and the Child Labor Laws: Massachusetts Employers' Guide*, a brochure/poster summarizing State and Federal laws for employers and explaining how employers can take action to prevent injuries to teens in their workplaces.

✓ **Job fairs:** The University of Massachusetts at Lowell mini-grant project distributed workplace safety information to employers participating in a youth "hiring hall"— a job fair for young people seeking summer jobs (and employers seeking summer help). The project also distributed material to youths, ensuring that both employers and youths received consistent and accurate information.

✓ **Small business development centers:** The Massachusetts State team partners with the Massachusetts Small Business Development Center Network to reach small business owners and managers who come into the centers for information about child labor laws. Every State has a similar network designed to provide technical assistance to small businesses.

Meeting the Challenge: Involving Employers

Involving employers in teen workers' safety is essential since they have primary responsibility for creating and maintaining safe workplaces. But involving employers can be especially challenging. Including OSHA compliance assistance specialists (CAS) on your team can help meet this challenge. Compliance assistance specialists are present in every OSHA Area Office. They provide outreach, training, education, and compliance assistance to businesses, business associations, labor unions, and other organizations.

Ruth McCully, former OSHA Region I administrator, and Douglas Edwards, Regional Compliance Assistance Coordinator, encouraged the CASs in their region to participate in young worker safety efforts. As a result, the CASs became active participants on many of the State teams. Several enrolled in the training-of-trainers workshops and now use activities from the young worker safety curricula when they are invited to schools. The specialists also include information about young worker safety at the workshops and educational forums they sponsor for employers.

The Region I office also discovered that employers do not have time to attend half-day workshops, but are willing to attend hour-and-a-half "donuts and coffee" meetings in the morning. This discovery led to a successful breakfast series about worker and young worker safety for small businesses.

To locate the compliance assistance specialist in your area, contact OSHA at (800) 321–6742 or visit the OSHA Web site at www.osha.gov. To learn more about the efforts in Region I, contact Douglas Edwards at (617) 565–2770.

Educating Health Care Providers

Health care providers, who are concerned with the health and safety of young people, as well as the costs of health care, can be especially useful in reaching business and industry groups. The teams in the northeastern States identified several opportunities to work with health care providers.

Massachusetts developed and disseminates to physicians and nurses *Protecting Working Teens: A Guide for Health Care Providers*. This brochure helps health care providers learn about the risks faced by young people in the workplace and gives them guidance on counseling their young working patients.

With assistance from EDC, this publication was adapted for use in Connecticut. The Connecticut team is concentrating its outreach on school-based health centers.

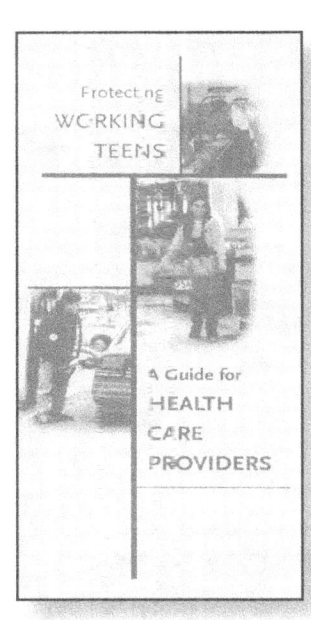

Health care provider guide (Massachusetts)

Using Data

State-wide or industry-specific occupational injury data is a powerful way of demonstrating the risks faced by young workers to teens, educators, parents, health care providers, and policymakers in your State. Collecting this data is often easier than it might appear, especially since some members of the State team have access to data collected by their agencies on a regular basis. Once collected and analyzed, this information can be used to customize curricula, brochures, fact sheets, and other educational material or to produce reports and press releases. It also is useful in evaluating the effect of policy or regulatory changes on injury rates. Many types of data are potentially available:

✓ **Hospital Data:** The State team member from New Hampshire's Department of Health and Human Services Injury Prevention Program searched hospital inpatient and emergency department data for injuries to teenagers whose expected payer source was the workers' compensation program. The data was included in an Injury Prevention Program report in order to focus the attention of the public and policymakers on the problem. The data was also examined to determine long-term trends, including the effect of eliminating work permits on injuries to teens in the workplace.

✓ **Employer's First Report of Occupational Injury or Disease:** The State team member from the New Hampshire Department of Labor's Safety and Training Division asked the State Workers' Compensation Division to send her copies of all Employer's First Report of Occupational Injury or Disease forms documenting injuries involving teenagers. This information is entered into a database and will be combined with hospital data collected by the Injury Prevention Program (see previous page).

✓ **Youth Risk Behavior Survey (YRBS):** YRBS, a national survey coordinated by the Centers for Disease Control and Prevention, includes a sample of high school students in grades 9–12. States can add a limited number of questions to the survey used in their State. Several of the Northeastern States are hoping to add occupational safety questions to their State's survey. The YRBS can generate data from a large sample with a rigorous methodology. It also has an established credibility with public health professionals, educators, and journalists.

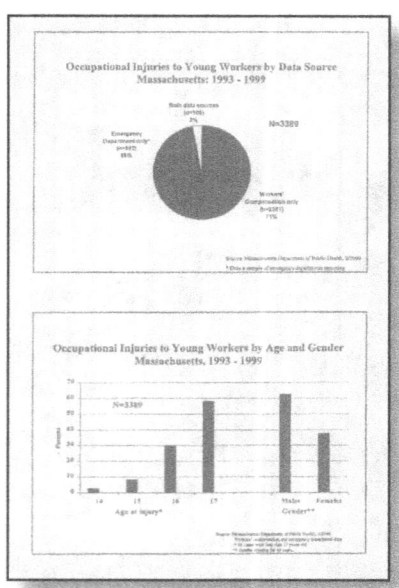

Young worker injury report (Massachusetts)

Young worker injury report (New Hampshire)

✓ **Surveillance System for Occupational Injuries to Youth:** The Massachusetts Department of Public Health's Teens at Work Injury Surveillance and Prevention Project uses multiple data sources to identify work-related injuries in youths. These include workers' compensation and emergency department data, hospital discharge reports, reports from individual physicians (in Massachusetts, hospitals and physicians are required to report occupational injuries to people under the age of 18), Massachusetts Burn Registry data, and data from the Census of Fatal Occupational Injuries and the Fatality Assessment and Control Evaluation. Cases that have sustained serious injuries or are indicative of broader problems are designated for followup interviews. Project staff conducts these interviews to learn about workplace hazards, safety training, adult supervision at work, and the impact injuries have on teens. Cases may be referred to other agencies if it appears that safety and health or child labor laws have been violated. The Department of Public Health may also conduct its own research-oriented investigation of workplaces to learn more about how teens are injured at work and to target interventions. Summary data from the surveillance system are used to identify high-risk industries and occupations for broader-based prevention efforts. With funding from NIOSH, the Massachusetts Department of Public Health is creating a "how-to" guide for other States interested in strengthening or expanding data collection systems.

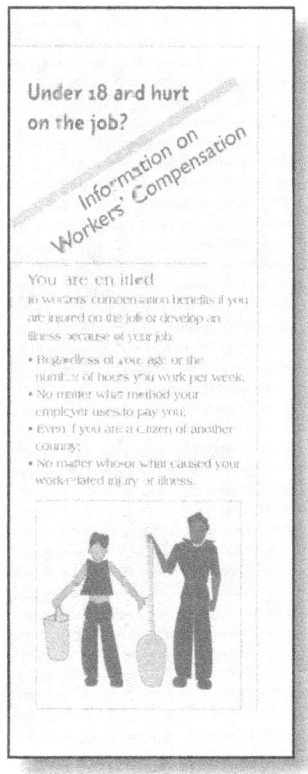

Teen worker's compensation guide (Massachusetts)

Providing Information about the Law

State and Federal laws and regulations govern the conditions of child labor. They restrict the types of jobs young people can hold, the kinds of machinery and materials they can use, the tasks in which they can engage, and the number of hours and times of day they can work. When Federal and State laws differ, the more restrictive law takes precedence.

State team members who are able to do so are involved in several types of information activities about laws:

✓ Providing information to States about laws, such as the fact that there are no Federal regulations about the hours 16 and 17 year olds are allowed to work

- ✓ Providing information about occupations and tasks that are now recognized as hazardous (e.g., using chemicals) or that have become more common for young people (e.g., repetitive motions)

- ✓ Promoting enforcement of existing laws

Examples of team activities include the following:

- ✓ **Tracking legislation:** Two bills introduced in the Connecticut legislature would have expanded the types of industries in which younger teens and preteens could be employed, as well as the times of year and hours they could work. Without the participation of the Connecticut Department of Labor, whose government relations staff track bills submitted to the legislature, the State team might not have been aware of this effort to change the law.

- ✓ **Responding to changes in the law:** The New Hampshire legislature abolished that State's work permit system with little warning or fanfare during the summer, when schools were not in session. Many educators, public health professionals, and occupational injury experts learned about this legislative action after it had been accomplished. Team members prepared information materials about the change.

- ✓ **Upgrading laws:** With prompting from the National Child Labor Coalition, the Maine Department of Labor updated that State's child labor laws. The list of occupations in which minors cannot work was expanded to include all occupations in Maine that are prohibited to minors under Federal law and a number of additional occupations, including those involving exposure to pesticides. The law was also strengthened in other ways, for example, by prohibiting those under the age of 18 from driving on the job or working alone in a cash business (in which there is a substantial risk of robbery).

In 1999, a bill enhancing the protection of young workers under State labor laws was submitted to the Massachusetts legislature. Only opponents testified at the legislative hearing, and the bill did not pass. It was then resubmitted in the next legislative session. The Massachusetts State team provided expertise on specific provisions of the bill to legislators and the public. The chair of the Accident Prevention and Poison Control Committee of the Massachusetts Chapter of the American Academy of Pediatrics (MCAAP) arranged to have members of the team brief the MCAAP about the bill.

Evaluation and Program Monitoring

Evaluation can provide an indication of a State team's impact and help the team refine its efforts and become more effective. The primary goal of a team is to increase the safety of teen workers. The most direct evidence of success in meeting this goal would be a reduction in occupational injury rates. However, identifying such a trend can be difficult. The availability, quality, and timeliness of data vary among States. And the number of detectable injuries emerging in the data systems of States with small populations could be relatively small, making it difficult to assess trends over a short time span. In fact, the increasing attention paid to occupational injuries to teens as a result of the work of the State teams could improve reporting of such incidents and result in an apparent increase in the injury rate.

A more realistic measure of success is the extent to which teams implement activities that are likely to promote young worker safety. Such activities might include educating teens, parents, educators, and employers; collecting and analyzing data on injuries and labor law violations; strengthening child labor laws; and improving the safety of workplaces. This publication highlights many of these successes.

State teams also investigated the impact of activities. For example:

- ✓ Pre- and post-tests were given to students as part of occupational safety classes to evaluate the overall increase in their knowledge and attitudes about occupational health and safety. When necessary, the names of students were removed from the forms before such analysis to preserve confidentiality.

- ✓ Evaluations completed by participants in employer, health care provider, parent, and educator symposiums were used to assess their reactions to the symposium, assess their needs, and plan future events.

- ✓ Pre- and post-tests for teachers trained in the use of a curriculum were used to measure both the knowledge gained from the training and their overall satisfaction with the training.

- ✓ State team activity logs are used to track a team's process and impact. (An activity log is a list of meetings, presentations, training, and other events; team members involved, the nature and size of the audience, and other information useful in assessing the team's value. Logs are also used to create capability statements for funding opportunities and justifying members' participation to their agencies.)

Systematic Reform

While the Massachusetts State team was far ahead of many States in looking at young worker injury data and defining research-based solutions to its injury problems, the team members came to the conclusion that they needed to institutionalize their efforts if they wanted to have a permanent impact on young worker safety in that State. In Massachusetts, the Departments of Labor, Education, and Health, as well as the Attorney General's Office, had some responsibility for the safety of young workers. But no established infrastructure was responsible for this issue. The demand for materials, information, training, and other resources by employers, schools, health care providers, parents, and youth was rising—yet no central office or agency could provide these services.

The Massachusetts State team modeled its efforts on work done in California. Guided by the recommendations developed by the California Study Group on Young Workers' Health and Safety, and using planning meetings based on the model developed in California, the Massachusetts State team evolved into the Massachusetts Young Worker Initiative (MYWI) and identified five goals for itself:

- ✔ Developing a systematic approach to educating teens, parents, employers, educators, and other professionals who work with youth about the labor laws and workplace health and safety

- ✔ Creating a central body to coordinate this educational effort and collect data on teen occupational safety and health

- ✔ Providing information about the Commonwealth's child labor laws

- ✔ Enhancing coordination among government agencies around this issue

- ✔ Fostering research on young workers, with an emphasis on occupational safety and health

Team representatives from MassCOSH, the University of Massachusetts at Lowell, and the Northeast Young Worker Resource Center provided essential logistical support. MYWI's first step was to expand its membership to include a wider range of stakeholders, including representatives from unions, health care organizations, businesses, and youth development organizations.

The members of MYWI realized that child labors laws alone could not protect young workers. Thus, they decided to focus on three additional areas needed to provide the State with a comprehensive young worker safety strategy:

- ✔ School-based strategies, including health and safety training for teens, involving school administrators and staff, and better using the work permit process to protect teen workers

- ✔ Work-based strategies, including on-the-job training for youth, improving the work environment, facilitating employer sharing of best practices, and providing information about child labor laws to employers

- ✓ Public information and community-based initiatives, including efforts involving health care providers, parent groups, and youth-serving agencies, media campaigns, and integrating occupational safety into community job training programs

MYWI working groups are developing recommendations in all of these areas for presentation to the legislature and the constituencies addressed in the report. They are also taking preliminary steps to create a young worker safety resource center as a single point of information and resources on this issue. Although this effort is still relatively new, its combination of using data and research, involving a broad coalition of public and private partners, developing a comprehensive State-wide strategy, and focusing on coordinating and institutionalizing young worker safety efforts may be a model of effecting permanent improvements in this area.

Resources

Young Worker Safety Resource Center

For general information about the Northeast Young Worker Network, specific information on activities in each State, and ordering information for all materials described in this guide, contact the YWSRC.

The Young Worker Safety Resource Center is funded by the Occupational Safety and Health Administration to serve State and local staff from job readiness programs, employers of youth, and other education and employment-related organizations that serve youth. The Center provides training on teaching teens about occupational safety; seminars for employers of youths; consultation and referrals for State departments of education and workforce development, as well as other organizations who want to conduct young worker safety training; and materials (described under "Materials" below). The Young Worker Safety Resource Center has two offices.

Education Development Center, Inc.

55 Chapel Street
Newton, MA 02458-1060

www.edc.org/

Chris Miara, Project Coordinator
(617) 618-2238

cmiara@edc.org

Labor Occupational Health Program

Center for Occupational and Environmental Health
University of California
2223 Fulton Street–4th Floor
Berkeley, CA 94720-5120

socrates.berkeley.edu/~lohp/

Diane Bush, Project Coordinator
(510) 642-5507

dbush@uclink4.berkeley.edu

State and Regional Contacts

These agencies and organizations are central to the issue of young worker safety and essential participants on a State team. Each description is followed by a contact that can help you locate agencies and organizations in your State and region.

Labor Department Contacts

Occupational Safety and Health Administration

OSHA develops and enforces safety and health regulations for all workers. OSHA staff have the expertise and resources to help State efforts promote the safety of teen workers.

OSHA has ten regional offices, with staff assigned to each State. Its Compliance Assistance Specialists are available to provide tailored information and training to employers and employees.

With the encouragement of OSHA, many States have developed and operate their own job safety and health programs, referred to as State Plans. OSHA approves and monitors State plans and provides up to 50 percent of an approved plan's operating costs.

There are currently 23 States and jurisdictions operating complete State Plans (covering both the private sector and State and local government employees) and three—Connecticut, New Jersey and New York—which cover public employees only.

For information and staff responsible for your State, contact:

> U.S. Department of Labor
> Occupational Safety and Health Administration (OSHA)
> 200 Constitution Avenue, N.W.
> Washington, DC 20210
>
> (800) 321-6742
> www.OSHA.gov

Wage and Hour Offices

Youths 14–17 years old are subject to Federal and State child labor laws. Therefore, staff from both Federal and State wage and hour offices are essential members of State teams.

The Federal Wage and Hour Division (WHD) is the office within the Federal Department of Labor that enforces Federal minimum wage, overtime pay, record-keeping, and child labor laws. There are five WHD regional offices and at least one

WHD office in every State. The WHD Web site provides information about these issues and includes a section on Federal child labor rules.

State Wage and Hour offices, typically located within State departments of labor, are responsible for enforcing State minimum wage, overtime pay, recordkeeping, and child labor laws.

For the Federal WHD staff responsible for your State, contact

>Toll free: (866) 487-9243
>www.dol.gov/esa/contacts/whd/america2.htm

For your State Wage and Hour Office, see

Interstate Labor Standards Association at www.ilsa.net.

Workforce Investment Boards (WIB)

Workforce Investment Boards oversee job training programs, including those for youth, funded through the Workforce Investment Act (WIA), a major Federal job training program. States are divided into WIA Service Delivery Areas, each of which is served by a WIB. These activities are coordinated at the State level. The control that WIBs have over funding and conditions at job training programs make them extremely beneficial partners on a State team.

For information on the Workforce Investment Act, contact

>Office of Career Transition Assistance
>Employment and Training Administration
>200 Constitution Avenue, NW–Room S4231
>Washington, DC 20210
>
>(202) 693-3045
>AskWIA@doleta.gov

For a list of the State offices coordinating WIA activities, see

>www.usworkforce.org/statecon.htm.

For an updated list of contact information for all Workforce Investment Act service delivery areas in the country, contact

>National Association of Counties
>440 First Street, NW–Suite #800
>Washington, DC 20001
>
>(202) 393-6226
>www.naco.org/programs/social/work/getstate.cfm.

Public Health Offices

Public Health Injury Prevention Program

Each State health officer designates an injury prevention director. The injury prevention director oversees injury prevention activities within the State health department and can identify other injury prevention personnel within the department (for example, in the State Maternal and Child Health Office).

To locate the injury prevention director in your State or territory, contact

> David Scharf, Executive Director
> State & Territorial Injury Prevention Directors' Association (STIPDA)
> 2965 Flowers Road South, Suite 105
> Atlanta, GA 30034
>
> (770) 690–9000
> www.stipda.org/

Maternal and Child Health (MCH) Office

Most State MCH offices have a designated MCH injury prevention coordinator. Many State health departments also have a designated adolescent health coordinator who addresses issues of adolescent health, including alcohol and other drug use and safe driving.

State contacts are available from the Federal MCHB or its national resource center, the Children's Safety Network

> Maternal and Child Health Bureau
> Injury and Violence Prevention Program
> Health Resources and Services Administration
> U.S. Department of Health and Human Services
> 5600 Fishers Lane–Room 18–A–38
> Rockville, MD 20857
>
> (301) 443–5720
> www.mchb.hrsa.gov/

> CSN National Injury and Violence Prevention Resource Center
> Education Development Center, Inc.
> 55 Chapel Street
> Newton, MA 02458–1060
>
> (617) 969–7100
> csn@edc.org
> www.childrenssafetynetwork.org

Occupational Health Programs

State occupational health programs conduct surveillance and educational activities to improve worker safety and health. They help employers comply with occupational safety and health laws, and operate laboratories to analyze potentially toxic substances found in current or former workplaces. NIOSH maintains a Directory of Occupational Safety and Health Contacts for State and Territorial Health Departments.

For a list of State contacts, see

> www.cdc.gov/niosh/statosh.html

Education Agencies

School-to-Work (STW)

STW offices coordinate activities designed to prepare youth for entry into the workplace. These activities typically involve local partnerships in which students are exposed to a variety of career choices through on-the-job experiences, apprenticeships, and mentoring. These offices (which are not always titled "School-to-Work") can be independent agencies, or located within various State departments, including education, commerce, and labor. To locate the STW office in your State, contact your State department of education.

Vocational Education Offices

State vocational education offices are responsible for vocational and technical education, which may take place in comprehensive high schools, vocational technical schools, vocational centers, and post-secondary educational institutions.

To locate the office in your State, contact

> National Association of State Directors of Vocational Technical Education
> The Hall of States
> 444 North Capitol Street, NW–Suite 830
> Washington, DC 20001
>
> (202) 737–0303
> www.careertech.org

Non-Governmental Organizations

American Academy of Pediatrics (AAP)

AAP has more than 55,000 members and at least one chapter in every State. AAP's Section on Injury and Poison Prevention (SIPP) is composed of members interested in preventing injuries for children. AAP has issued policy statements on *The Hazards of Child Labor, Prevention of Agricultural Injuries among Children and Adolescents,* and *The Role of the Pediatrician in Transitioning Children,* and *Adolescents with Developmental Disabilities and Chronic Illnesses from School to Work or College,* all of which can be found on the AAP Web site.

To locate the AAP chapter and SIPP members in your State, contact

> American Academy of Pediatrics (AAP)
> 141 Northwest Point Boulevard
> Elk Grove Village, IL 60007–1098
>
> (847) 434–4000
> www.aap.org

COSH (Committee for Occupational Safety and Health)

COSHes are coalitions of union representatives, workers, physicians, lawyers, and health and safety advocates who provide occupational safety and health training to unions, workers, and young people, and educate the public about occupational health and safety. There are about 25 COSHes nationwide as well as a number of COSH-related organizations. A list of these can be found on the NYCOSH Web site at www.nycosh.org/link-resources.htm/#anchor565960.

For more information on locating a COSH or affiliated organizations, visit

> www.coshnetwork.org

Safety Council

Most States have at least one Safety Council, local affiliates of the National Safety Council that provide training, conferences, workshops, consultation, newsletters, legislative, regulatory, and research updates and safety materials at the State and community level. The National Safety Council began with a focus on occupational safety but has since expanded its focus into other areas of injury prevention.

For information about State Safety Councils, contact

> National Safety Council
> 1121 Spring Lake Drive
> Itasca, IL 60143–3201
>
> (630) 285–1121
> www.nsc.org/chaptop.htm#LIST

Unions

Unions may be especially valuable for addressing safety in a particular industry in which they have a substantial membership. While not every union is an affiliate of the AFL-CIO, it is a good national resource for locating a union in your area that may be interested in working with you on the issue of young worker safety.

For local unions, contact

> AFL–CIO
> 815 16th Street, NW
> Washington, DC 20006
>
> (202) 637–5000
> www.aflcio.org

Other Sources of Information

National Institute for Occupational Safety and Health (NIOSH)

NIOSH is the Federal agency responsible for conducting research and making recommendations for preventing work-related illnesses and injuries. NIOSH collects and analyzes occupational health data; conducts investigations and evaluates hazardous working conditions, chemicals, and machinery; and develops and disseminates information about preventing occupational diseases, injuries, and disabilities.

NIOSH funds the regional Centers for Agricultural Disease and Injury Research, Education, and Prevention on issues that include child agricultural health and safety. NIOSH also supports the work of the National Children's Center for Rural and Agricultural Health and Safety (NCCRAHS), located in Marshfield, Wisconsin. NCCRAHS engages in research and prevention activities on agricultural safety and health as well as other health and safety issues pertinent to children living in rural areas.

For general information about NIOSH, contact

> (800) 356–4674
> www.cdc.gov/niosh/

For information about the regional Centers for Agricultural Disease and Injury Research, Education, and Prevention:

> www.cdc.gov/niosh/agctrhom.html

For information about NCCRAHS

> Toll free phone: (888) 924–7233 or (715) 389–4999
> e-mail: nccrahs@mfldclin.edu
> www.research.marshfieldclinic.org/children

Bureau of Labor Statistics (BLS)

The BLS collects, analyzes, and disseminates statistical data on work and the conditions of work in the United States.

The BLS can be reached at

>   202–606–5886
>   www.bls.gov

Youth Rules

A Web site created by the U.S. Department of Labor with resources for parents, teachers, teens, and employers.

>   www.youthrules.dol.gov

Youth 2 Work

A Web site created by OSHA with teen worker safety and health information.

>   www.osha.gov/SLTC/teenworkers/index.html

Resource Materials

Young Worker Safety Resources is an annotated list of materials, including those described in this guide. This list can be obtained from

>   The Young Worker Safety Resource Center
>   Education Development Center, Inc.
>   55 Chapel Street
>   Newton, MA 02458–1060
>
>   (617) 618–2238
>   cmiara@edc.org
>   www.edc.org/

The following are items that can be particularly useful in your State efforts to protect the safety of young workers.

Curricula and Teachers' Guides

Safe Jobs for Youth: A Theme-Based Curriculum Unit for High School Students
UCLA Labor Occupational Safety and Health Program, 2000

This 10-class, 2-week curriculum is designed to give young people information and skills related to workplace safety and health. This material uses interactive, student-centered activities. The lesson plans cover a variety of topics including child labor law information, job safety hazards and solutions, handling sexual harassment on the job, and workers' compensation for working teens. It is designed for the 9th grade, but is also very appropriate for 10–12 grade students. Includes the 12-minute video, *Your Work–Keepin' It Safe,* which covers safety and health hazards in fast food, construction, and grocery stores and shows teens teaching teens. The video can also be ordered separately.

Contact: UCLA-LOSH Program, 6350B Public Policy Bldg., Los Angeles, CA 90095–1478, (310) 794–5964. Available to view or download online at www.sppsr.ucla.edu/res_ctrs/iir/losh/

Safe Jobs, Safe Youth: A Teacher's Resource Kit
California Young Worker Resource Network, 2000

This packet includes a plan for teaching students about basic safety and legal rights on the job, a short interactive activity that offers an introduction to the many health and safety issues that employment raises for youth, a *Safe Jobs for Youth* poster, a copy of the pamphlet, *Are You a Working Teen?*, stickers, and an order form for requesting further working teen resources.

Contact: LOHP, University of California at Berkeley, 2223 Fulton St., Berkeley, CA 94720–5120, (510) 642–5507 or download at www.youngworkers.homestead.com.

Safe Work/Safe Workers: A Guide for Teaching High-School Students about Occupational Safety and Health, Massachusetts Department of Public Health and Education Development Center, rev. 2001

This 3-hour curriculum uses interactive activities to teach teens about workplace hazards, effective strategies to prevent occupational illnesses and injuries, their rights on the job, and the resources available to assist them. Includes a 10-minute video entitled *Teens: The Hazards We Face in the Workplace,* which includes interviews by teens of other teens who were injured at work.

Contact: Children's Safety Network, EDC, 55 Chapel Street, Newton, MA 02458, (617) 618–2207

Starting Safely: Teaching Youth about Workplace Safety and Health
Workplace Development Center, 2000

This teacher's guide is designed to teach high school students the basic concepts of occupational health and safety and to raise their awareness about these issues. It uses a video and a series of interactive activities.

Contact: Maine Department of Labor, Bureau of Labor Standards, 45 State House Station, Augusta, ME 04333–0045, (207) 624–6400.

Work Safe! A Health And Safety Curriculum For Youth Employment Programs
Labor Occupational Health Program, 2000

This curriculum is designed to help job training programs teach their youth participants about job health and safety in a fun and interesting way. It consists of four learning activities and includes handouts and a copy of *Are You a Working Teen? Protect Your Health, Know Your Rights.*

Contact: LOHP, University of California at Berkeley, 2223 Fulton Street, 4th floor, Berkeley, CA 94720–5120, (510) 642–5507

Educational Resources for Teens, Employers and Parents

Are You a Working Teen? What You Should Know About Safety and Health on the Job
National Institute for Occupational Safety and Health, 1997

This brochure gives information to teens about the dangers of occupational injury and the rights of teens in the workplace. It also outlines the Federal laws and regulations regarding adolescent employment, specifically what types of jobs and the number of hours teens are allowed to work. It provides resources on workplace safety and rights.

Contact: NIOSH at 1–800–35–NIOSH or www.cdc.gov/niosh/adoldoc.html

Tools for Orienting Work Site Supervisors about Teen Health and Safety
Labor Occupational Health Program, 2000

This is an information packet for work site supervisors, with four tools to use in job training programs:

(1) Checklist for Job Trainers and Job Developers

(2) Safety Training Agreement

(3) Safety Orientation Checklist

(4) Facts for Employers—Safe Jobs for Teens

Contact: LOHP, University of California, 2223 Fulton Street, Berkeley, CA 94720, (510) 642–5507

Parents' Primer: When Your Teen Works
National Consumers League, 1997

This pamphlet advises parents on the number of hours teenagers of different ages should work; how to prevent teens from becoming involved in hazardous employment; warning signs that a teen is working too much or in a hazardous workplace; and information about Federal child labor laws.

Contact: National Consumers League, 1701 K Street, N.W., #1201, Washington DC 20006, (202) 835-3323 or view online at www.stopchildlabor.org/aboutus/clcresources.htm

Reports and Guides for Professionals

Promoting Safe Work for Young Workers: A Community-Based Approach
National Institute for Occupational Safety and Health (NIOSH), 1999

This resource guide documents the experiences of three projects funded by NIOSH to promote safety and health for young workers. The guide provides summaries of the three projects, gives facts about young worker safety and health, and lists steps in coordinating a young worker project. Detailed guidance is given for working with schools, employers, parents, health care providers, job training programs and teen peer education programs.

Contact: NIOSH, Publications Dissemination, 4676 Columbia Parkway, Mail Stop C-13, Cincinnati, OH 45226–1998, (800) 35-NIOSH or e-mail: pubstaft@cdc.gov

Protecting Youth at Work: Health, Safety and Development of Working Children and Adolescents in the United States
Committee on the Health and Safety Implications of Child Labor
Institute of Medicine, 1998

The committee presents a wide range of data and analysis on: the scope of youth employment; factors that put children and adolescents at risk in the workplace; and the effects of employment on health, educational attainment and lifestyle choices. The committee recommends specific initiatives for legislators, regulators, researchers and employers.

Contact: National Academy Press (800) 624–6242 or (202) 334–3313 or order on-line at www.nap.edu/bookstore

Report on the Youth Labor Force
U.S. Department of Labor, 2000

Presents a brief summary of key aspects of the U.S. laws and regulations governing child labor. Provides a detailed look at youth labor in this country, including how it differs among major demographic groups, between agricultural and non-agricultural sectors and overtime. Describes the outcomes of young people's work activities, including occupational injuries and fatalities and other long-term consequences.

Contact: US DOL, Available for download at www.bls.gov/opub/rylfhome.htm

www.ingramcontent.com/pod-product-compliance
Lightning Source LLC
Chambersburg PA
CBHW081752170526
45167CB00009B/4010